AMAZING SIGHTS
RAINBOWS

by Jane P. Gardner

pogo

Ideas for Parents and Teachers

Pogo Books let children practice reading informational text while introducing them to nonfiction features such as headings, labels, sidebars, maps, and diagrams, as well as a table of contents, glossary, and index.

Carefully leveled text with a strong photo match offers early fluent readers the support they need to succeed.

Before Reading

- "Walk" through the book and point out the various nonfiction features. Ask the student what purpose each feature serves.
- Look at the glossary together. Read and discuss the words.

Read the Book

- Have the child read the book independently.
- Invite him or her to list questions that arise from reading.

After Reading

- Discuss the child's questions. Talk about how he or she might find answers to those questions.
- Prompt the child to think more. Ask: Have you ever seen a rainbow? Have you ever wondered how rainbows form?

Pogo Books are published by Jump!
5357 Penn Avenue South
Minneapolis, MN 55419
www.jumplibrary.com

Copyright © 2021 Jump!
International copyright reserved in all countries. No part of this book may be reproduced in any form without written permission from the publisher.

Library of Congress Cataloging-in-Publication Data

Names: Gardner, Jane P., author.
Title: Rainbows / Jane P. Gardner.
Description: Minneapolis, MN: Pogo Books, [2021]
Series: Amazing sights in the sky | Includes index.
Audience: Ages 7-10 | Audience: Grades 2-3
Identifiers: LCCN 2019059862 (print)
LCCN 2019059863 (ebook)
ISBN 9781645275688 (hardcover)
ISBN 9781645275695 (paperback)
ISBN 9781645275701 (ebook)
Subjects: LCSH: Rainbows–Juvenile literature.
Classification: LCC QC976.R2 G37 2021 (print)
LCC QC976.R2 (ebook) | DDC 551.56/7–dc23
LC record available at https://lccn.loc.gov/2019059862
LC ebook record available at https://lccn.loc.gov/2019059863

Editor: Jenna Gleisner
Designer: Molly Ballanger

Photo Credits: orxy/Shutterstock, cover; Lisa-S/Shutterstock, 1; StockImageFactory.com, 3; patpitchaya/Shutterstock, 4; Francisco Blanco/Shutterstock, 5; Michael Warwick/Shutterstock, 6-7; Mila Drumeva/Shutterstock, 8-9; dhughes9/iStock, 10-11; James Wheeler/Shutterstock, 12-13; Veni vidi...shoot/iStock, 14-15; Fexel/Shutterstock, 16-17 (background); Anna Nahabed/Shutterstock, 16-17 (girl); Richard Wayman/Alamy, 18; Frank Olsen, Norway/Getty, 19; kuruneko/Shutterstock, 20-21; Menno van der Haven/Shutterstock, 23.

Printed in the United States of America at Corporate Graphics in North Mankato, Minnesota.

TABLE OF CONTENTS

CHAPTER 1
Water and Light 4

CHAPTER 2
Viewing Rainbows 12

CHAPTER 3
Rainbows Everywhere 18

ACTIVITIES & TOOLS
Try This! 22
Glossary 23
Index 24
To Learn More 24

CHAPTER 1
WATER AND LIGHT

It rained all morning. As the rain lets up, the clouds part. The sun starts to shine. A rainbow arches across the sky.

Rainbows often form after rain. They can show up in fog, too!

fogbow

CHAPTER 1

Some appear near waterfalls. They can even form from the spray of a garden hose. Why? What do all of these have in common? Water! Rainbows need raindrops or water to form. They also need sunlight.

DID YOU KNOW?

Millions of rays of sunlight go into making one rainbow!

Rainbow Falls, Hawaii

prism

spectrum

Light is **energy** that travels through space. All the light we see is **white light**. White light is made up of a **spectrum** of colors. Red, orange, yellow, green, blue, indigo, and violet make up the spectrum.

Water acts like a **prism**. Light **refracts**, or bends, when it passes through a prism or water drops. The light bends and splits into the spectrum of colors.

CHAPTER 1

TAKE A LOOK!

Light bends as it enters a raindrop. Some light **reflects** off the inside of the drop. It bends again as it leaves. This is how a rainbow is made.

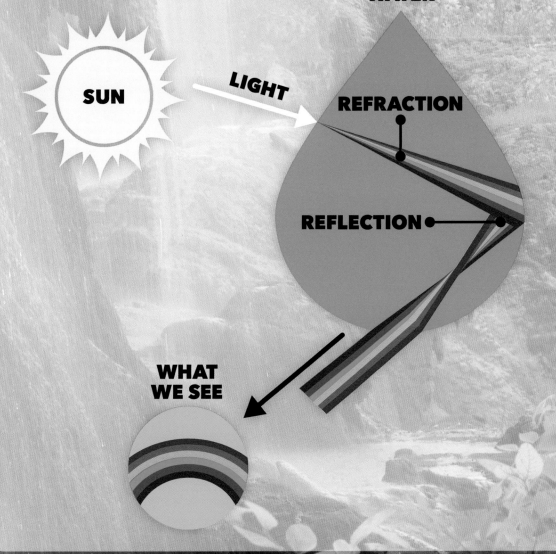

CHAPTER 2
VIEWING RAINBOWS

Have you ever seen a double rainbow? This happens when light is reflected twice in the raindrops. The result is two rainbows!

Even though it is usually faint, there is always a second rainbow just above the first! Look carefully. One is paler than the other. In the brighter rainbow, red is at the top of the arch. In the paler rainbow, red is at the bottom.

Rainbows are **optical illusions**. We see them. But they aren't physically there. You cannot touch them, and there is no pot of gold at the end. How they look depends on where we are and where the sun is.

DID YOU KNOW?

Rainbows are actually full circles. But we can only see what is above the **horizon**. That is why we see rainbows as arches.

The sun has to be behind you to see a rainbow. The water has to be in front of you. For this reason, rainbows are more common in the early morning and late afternoon. Why? The sun is lower in the sky at these times. Instead of shining down on you, it is behind you.

> **DID YOU KNOW?**
>
> Why do rainbows appear so large? Each time light refracts, the bands of color get wider and wider.

CHAPTER 2

CHAPTER 3
RAINBOWS EVERYWHERE

The next time it rains, look for the sun. Then look in the opposite direction. Triple and even quadruple rainbows are possible!

Moonbows can appear at night. The light comes from sunlight reflected off the moon's surface instead of direct sunlight. Moonbows are usually less bright than rainbows because there is less light involved. They appear opposite the moon.

CHAPTER 3 19

If it doesn't rain, make your own rainbow. Remember, all you need is light and water!

CHAPTER 3   21

ACTIVITIES & TOOLS

TRY THIS!

MAKE A RAINBOW

You don't have to wait for a rainy day to see a rainbow. Make your own rainbow indoors with this activity!

What You Need:
- clear glass of water
- sunshine
- one sheet of white paper
- spray bottle
- window

❶ Place a clear glass of water on the floor in the sunshine.

❷ Place the paper flat on the floor on the opposite side of the glass from the incoming sun's rays.

❸ Spray water on the window. Watch as the rainbow appears on the piece of paper!

GLOSSARY

energy: Power from a source that produces light and heat.

horizon: The line where Earth seems to meet the sky.

optical illusions: Visions we see but that do not actually exist.

prism: A clear, solid object with several flat sides called faces.

reflects: Throws back light from a surface.

refracts: Bends.

spectrum: A band of colors formed when a beam of light passes through a prism or water droplet.

white light: The mixture of all of the colors of the spectrum. The light we see is white light.

INDEX

afternoon 16
clouds 4
double rainbow 12
fog 5
horizon 15
light 9, 10, 11, 12, 16, 19, 21
moon 19
moonbows 19
morning 4, 16
optical illusions 15
prism 10
quadruple rainbows 18

rain 4, 5, 18, 21
raindrops 6, 11, 12
rays 6
reflects 11, 12, 19
refracts 10, 11, 16
spectrum 9, 10
sun 4, 11, 15, 16, 18
sunlight 6, 19
triple rainbows 18
water 6, 10, 11, 16, 21
waterfalls 6
white light 9

TO LEARN MORE

Finding more information is as easy as 1, 2, 3.

❶ Go to www.factsurfer.com
❷ Enter "rainbows" into the search box.
❸ Click the "Surf" button to see a list of websites.